Rain

Written by Andrew Kelly
Illustrated by Caroline Campbell

Where does rain come from?

How does the beach change from this . . .

to this?

As strange as it may seem,
the things that make rain
are all in this picture.

Rain begins with something very hot—the sun.

The sun heats the water on the surface of the sea, and changes it into a gas.

This gas is called vapor.

You can't see this happening, because vapor is invisible.

But if you put a tray of water
outside in the sun
and leave it for several hours,
all the water will be gone
when you come back.

The sun will have turned
the water in the tray to vapor.

This is what the sun does to the water
on the top of the sea.

Vapor is very, very light,
so it floats up and up.

The higher it goes,
the colder it becomes.
The vapor becomes so cold
that it changes into tiny drops of water.

The drops of water are so small
that they, too, float up.

As they go up, they get bigger and bigger,
because two drops become
one bigger drop,
and two bigger drops
become a very big drop.

The drops eventually become so big that, if you were close enough, you could see them.

You cannot see each drop as it is high in the sky, but you can see them all together— as a cloud.

Clouds are lots of drops of water floating together in the air.

The clouds rise higher and higher, and the drops keep getting bigger and bigger.

In the end, the drops inside the cloud
become so big
that they are too heavy to float,
and they fall out of the sky,
as . . .

RAIN!